Bu

Prepare for the API Economy

Kin Lane
&
Audrey Watters

Version

/2011/05/

Access to the Web is now a human right.

It's possible to live without the Web. It's not possible to live without water. But if you've got water, then the difference between somebody who is connected to the Web and is part of the information society, and someone who (is not) is growing bigger and bigger.

- Sir Tim Berners-Lee

Table of Contents

1. *Introduction*
2. *What is an API?*
3. *Who uses APIs?*
4. *What type of technologies go into an API?*
5. *History of modern Web APIs*
6. *Where some APIs stumble*
7. *Building blocks for a successful API*
8. *Fostering an API ecosystem*
9. *BizDev 2.5*

1
INTRODUCTION

The popular photo-sharing site Flickr launched in February 2004. Six months later, it released its API -- Application Programming Interface -- an interface that enabled users and other applications to connect and interact with the photos and the website. Six months after the API was released, Flickr was acquired by Yahoo.

The launch of the API helped Flickr quickly become the image platform of choice for the early blogging and social media movement as it gave users the ability to easily embed their Flickr photos into their blogs and social network streams.

But the API didn't simply foster more adoption and usage among individuals. It also provided the photo-sharing company with a new way to connect to other businesses. Indeed, Flickr co-founder Caterina Fake pointed to the API as a key way in which the company was able to have other third-party developers build out the platform and provide services that the photo-sharing site didn't yet offer.

Fake noted that, as is the case with many growing businesses, Flickr executives had little time in their schedules to arrange meetings to hash out partnerships with third parties. Instead Fake says, Flickr encouraged companies to "feel free to apply for a Commercial API key and build something off the API."

"Biz Dev 2.0," she called it.

Fake describes the traditional sorts of business development processes — "spending a lot of money on dry cleaning, animating your PowerPoint, drinking stale coffee in windowless conference rooms and scouring the thesaurus looking for synonyms for 'synergy'. Not to mention trying to get hopelessly overbooked people to return your email. And then after the deal was done, squabbling over who dealt with the customer service."[1] An API, she contends, can help companies bypass this. While a Web 1.0 world made partnerships and integration challenging, a Web 2.0 world has the advantage of the API, something that removes many of the bureaucratic, legal, and technical obstacles.

In other words, by developing and marketing your API, your company can tap into this Biz Dev 2.0 and can take advantage of new technologies and new business development practices.

The Purpose of this Book

Recent years have seen API adoption explode among developers, for reasons that this book will examine. But the purpose of this book is to help scale out the business side to meet this rising developer demand. Written by someone with an engineering and a business

---

[1] http://goo.gl/bPean

background, *The Business of APIs* also aims to bridge the technical and the business sides of API development.

This book aims to help people understand what APIs are and who uses them and to inform people about the different types of APIs that are available. As the title suggests, this is a business-oriented book. Nonetheless it does seek to educate users about what types of technologies go into popular Web APIs. The book also surveys the history of modern Web APIs and examines how they've been used successfully.

If you are considering launching an API, this book should help you understand the common stumbling blocks that have been faced by many API owners -- then hopefully you can avoid them. The book will also identify common building blocks used by API owners, building blocks that should be fundamental for your API planning and development.

The Business of APIs highlights what it takes to be successful in providing quality Web APIs and points to some of the innovative steps companies are taking with their APIs -- all in an effort to build vibrant API ecosystems and healthy businesses.

2
WHAT IS AN API?

An Application Programming Interface (API) is a way for websites, programs, and mobile applications to talk to each other. For example, APIs can allow access to a database, they can provide additional functionality not part of a program's existing specs, or they can control external gadgets.

Companies develop APIs for a number of reasons: to make resources available both internally and externally and to give partners and other developers access to their information, services and hardware.

APIs come in many shapes and sizes. The most common type is referred to as Representational State Transfer or REST. REST works just like any other Web page on the Internet, meaning that the communication of information facilitated by an API is passed as part of the data that loads in your browser.

In part, because these REST APIs use the same protocol as a webpage, they have become ubiquitous. You use APIs every time you use your credit card. Websites you visit get their information and functionality through APIs. And all the programs on your mobile phone depend on APIs. Of course, we don't often think about what occurs on the "back end," as user interfaces disguise this transfer of data. Nonetheless, APIs are everywhere.

3
WHO USES APIS?

APIs are launched primarily to give partners that are outside the company "firewall" access to data and resources. As technologies like REST simplify APIs, APIs in turn are being used by more and more of the general public, including non-developers.

So as APIs' ease-of-use and popularity increases -- and as APIs demonstrate their value and deliver efficiencies -- many companies have begun to consume their own APIs. These companies are building internal systems, websites, and mobile apps using the same APIs that they make available to third-party developers and to the public.

Used by Partners

Sharing information and services with your partners is vital to your business. With the complexities of businesses systems and security of firewalls, this exchange isn't always easy.

You rely on the relationships you've established with your partners to make your business thrive, and often showcase these relationships to the public to emphasize the strength of your business.

Defining what resources you should make available to partners -- and then making them available for each individual partner -- can be a costly and time-intensive endeavor.

While a Web API can't define this for you, it can certainly facilitate the sharing and delivery of resources to partners.

Used by Developers

One of the benefits of Web APIs as opposed to other, older API protocols is that they are understood by a wide variety of developers from freelancers to enterprise developers. That makes APIs a standard interface, of sorts, that all types of developers can understand, and an interface that can be used in all types of programming languages.

As these Web APIs are designed to make life easier for developers, then every API owner should focus on empowering developers from all business sectors -- that means a general audience of developers, not simply those within partner relationships. These outside developers can adopt APIs and innovate in ways that neither a company nor their partners could have imagined.

An API that is opened up to the public, and allows any developers to register and use is often called an Open API.

Used Internally

APIs may be built for external access, for partners and developers outside your company's firewall. But in larger companies, with multiple departments and divisions, for example, or spread across geographic regions, it can be difficult to get access to information and development resources too. So APIs can offer the same benefits internally as they do externally.

Used by the Public

API usage will likely move beyond the developer community in the coming years, just as having one's own website -- once something that only a programmer could do -- is now commonplace. There may be ways for you to build widgets, links, buttons and other embeddable tools that can be utilized by non-developers.

The point here is that you should constantly review both the actual users and the potential users of your API, and in doing so, keep in mind that an API can be valuable to a much wider group than originally imagined.

4
WHAT TYPE OF TECHNOLOGIES GO INTO AN API?

This book addresses the business side of APIs, but because APIs are driven by technology, it's important that anyone involved have at least a basic awareness of the technologies that go into delivering a web API.

APIs are driven by a set of specific technologies, making them easily understood by mainstream developers. This standardization means that APIs can work with common programming languages.

There are two major approaches to delivering APIs: REST and SOAP.

SOAP Protocol

SOAP, or Simple Object Access Protocol, pre-dates REST. It was originally designed in 1998 as a Microsoft project and became a W3C Recommendation in 2003. Even though SOAP has a long history with enterprise development, it has not seen the same adoption with the Web 2.0 developer community, due to its complex and verbose nature.

SOAP is far from dead, but in the new generation of web-based APIs, REST interfaces returning JSON are fast replacing the bulky SOAP interfaces that returns XML.

REST Standard

REST describes a specific standard for delivering an API. REST takes advantage of the same Internet mechanisms that are used to view regular Web pages. In other words, HTTP, the protocol we use to browse the Web, has built-in ways to transfer, describe and deliver content to humans. It makes sense to take advantage of this existing infrastructure to deliver data and functionality between applications as well.

There are different opinions of what it means to be truly REST compliant, but many developers have just agreed that using existing HTTP mechanisms for your API is considered "RESTful."

By using HTTP, REST does not need any additional servers or protocols to deliver or view REST APIs. The reduced complexity of REST makes it more efficient to use in development, and as such, has made it the preferred choice of developers, application architects, and API owners.

The Language of APIs

REST allows you to take data and functionality available on your website and make these resources available through a Web API. Then, instead of returning HTML to represent these resources, the API returns a XML or JSON version of the information.

XML

Some APIs return XML, Extensible Markup Language. XML was originally designed to describe documents, but has been adapted to describe anything from contacts to geographic data.

Much like HTML, XML uses opening and closing tags to describe content. A XML transmission describing the author of this book might look like:

```xml
<contact>
    <firstname>Kin</firstname>
    <lastname>Lane</lastname>
    <age>38</age>
    <website>apievangelist.com</website>
    <address>
        <streetaddress>123 2nd Street</streetaddress>
        <city>New York</city>
        <state>NY</state>
        <postalcode>10021</postalcode>
    </address>
</contact>
```

This Gist brought to you by GitHub.   Business of APIs - XML view raw

XML has long been the standard for Web API communication. However recently it has been losing ground to JSON because XML is considered more bloated and less efficient. As the popularity of REST increases, JSON may continue as the preferred format, but XML will still have a place with many older developers.

JSON

JavaScript Object Notation or JSON is a way for programs to exchange information, and is very similar to XML.

JSON uses brackets, quotes, colons and commas to separate data, and give the information some meaningful structure. A JSON description of the author might look like:

```
{
    "firstName": "Kin",
    "lastName": "Lane",
    "age": 38,
    "address":
    {
        "streetAddress": "21 2nd Street",
        "city": "New York",
        "state": "NY",
        "postalCode": "10021"
    }
}
```

This Gist brought to you by GitHub.   Business of APIs - JSON view raw

JSON is a light weight, simple way to exchange data across the Internet.

API AUTHENTICATION

APIs require authentication to use. In other words, any application that needs to send or receive data through an API needs to provide credentials to ensure it has authorized access.

There are two main types of authentication: Basic Auth and OAuth.

Basic Auth

Basic Auth is a way for a Web browser or application to provide credentials in the form of a username and password. Because Basic Auth is integrated into the HTTP protocol, it is the easiest way for users to authenticate with a REST API.

However while Basic Auth is easily integrated, if SSL (Secure Socket Layer) is not used, the username and password are passed in plain text and can be easily intercepted on the open Internet. That makes OAuth, described below, a much better choice for REST API authentication. Nonetheless, Basic Auth is perfectly suited for APIs that are intended for a wider audience and do not give access to sensitive information.

OAuth

OAuth is an open protocol to allow secure API authorization for web and desktop applications, mobile phones, and other devices. OAuth is widely considered the industry standard for API Authentication.

Virtually any application resource can be shared via an API, including photos, product, location or user information, and OAuth can issue access tokens for each of these individual resource areas. OAuth tokens can be unique per developer and for each API service they access, and as such provide granular security for all API services areas.

Final Thoughts on API Technologies

We are beginning to see now a standard way to deliver, a common language to communicate, and a secure way to authenticate with APIs. This didn't occur overnight, it has take the last ten years to define this approach.

To appreciate where we are with APIs today, it's necessary to explore and understand the history of Web APIs.

5
HISTORY OF MODERN WEB APIS

APIs have been around since the 1980s, when they were used in hardware and software development. However, the history of the modern Web API is fairly short -- just a little over ten years.

There are several pioneers of Web APIs, and while they didn't necessarily invent any of the technologies at play here, they did popularize their usage and establish some of the common practices. Many of these pioneers have shaped the way in which we develop, deploy, consume, and support APIs.

This chapter moves through a brief history of Web APIs, looking at how APIs have been instrumental in the development of several key sectors: e-commerce, mobile phone applications, cloud computing, and social networking.

First Mover: Salesforce

In February 2000, Salesforce officially launched its enterprise-class, web-based, sales force automation as an "Internet as a service."

An XML API was part of Salesforce from day one. Salesforce identified that customers needed to share data across their different business applications, and APIs were the way to do this.

Marc Benioff, Salesforce CEO and founder stated, "Salesforce is the first solution that truly leverages the Internet to offer the functionality of enterprise-class software at a mere fraction of the cost."

Salesforce was the first cloud provider to take an enterprise class web application and API and deliver what we know today as Software-as-a-Service (SAAS).

APIs and the Growth of E-Commerce

E-Commerce was one of the original promises of the Internet: the World Wide Web was going to change the way we do business. So opening up the sale and purchase of products to third parties just makes sense, and so it's no surprise that e-commerce is where APIs first began gaining acceptance.

eBay

In November 2000, eBay launched the eBay API, along with the eBay Developers Program. The eBay API was initially rolled out to only a select number of "licensed" eBay partners and developers.

As eBay stated, "Our new API has tremendous potential to revolutionize the way people do business on eBay and increase the amount of business transacted on the site, by openly providing the tools that developers need to create applications based on eBay technology, we believe eBay will eventually be tightly woven into many existing sites as well as future e-commerce ventures."

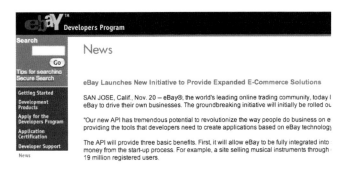

The eBay API was a response to the growing number of applications that were already relying on their site -- either legitimately or illegitimately. The official API aimed to standardize how applications integrated with eBay, making it easier for partners and developers to build a business around the eBay ecosystem.

Amazon

In July 2002, online retailer Amazon launched Amazon Web Services, allowing developers to incorporate Amazon content and features into their own websites. Amazon product data -- for search and display -- was accessible through XML and SOAP standards.[2]

---

[2] http://goo.gl/Sk8ko

From day one, the API was integrated with the Amazon Affiliate Program, allowing developers to monetize their sites through purchases made at Amazon via links from their websites.

Internet visionary Tim O'Reilly was cited in original Amazon Web Services press release, saying "This is a significant leap forward in the next-generation programmable Internet."

Web APIs might have started in e-commerce, but eventually these grew into new areas -- social media, cloud computing and mobile technology. But before doing so, the "dot com bubble" seemed to stagnate API development. And it would take about five years from these early eBay and Amazon APIs before the next wave of significant API growth.

The Web Gets Social (via APIs)

That next growth phase started in 2005 with Flickr, and soon spread into the social networking and social media world we know now, dominated by Facebook and Twitter.

Facebook

Facebook launched its long awaited development platform in February 2006.[3]

---

[3] http://goo.gl/rGEdk

Version 1.0 of the Facebook Development Platform gave developers access to Facebook friends, photos, events, and profile information.

Almost immediately, developers began to build social applications, games, and mashups on the development platform. This gave Facebook an edge over its competitor MySpace and helped establish Facebook as the top social platform, particularly with the phenomenal successes of games like Farmville.

Twitter

In September 2006, the micro-blogging service Twitter launched the Twitter API.[4] Much like the release of the eBay API, Twitter's API was a response to the growing usage of Twitter by those scraping the site or creating rogue APIs.

Twitter's API was REST, and In the beginning, Twitter used Basic Auth for authentication. However, four years later it announced a switch to OAuth, resulting in the "OAuth Apocalypse" when developers were given a countdown to update their apps or lose authentication and access.[5]

---

[4] http://goo.gl/s8WkL
[5] http://goo.gl/83dzG

Introducing the Twitter API
Wednesday, September 20, 2006

Some smart folks out there on the that series of tubes we call the Internet have been putting together interesting projects like this map and this page without any help from us so we thought it was high time to release an API.

Although Twitter has maintained a vibrant ecosystem, the "OAuth Apocalypse" has just been one area in which the company's development plans have run into conflict with third-party developers.

The API Mashup

As the social Web became more popular and as more data became available via APIs, a number of "mashups" were created, pulling data from multiple APIs and aggregating them -- "mashing them up" -- into a new service.

Undoubtedly, one of the most popular APIs for mashups is Google Maps. The ability to merge data with geographic location evolved the notion of what is possible with an API.

Google Maps

Google launched the Google Maps API in June 2006, allowing developers to put Google Maps data on their own sites using JavaScript.[6]

The API launch came just shy of six months after Google released Google Maps as an application. The API was, again, in direct response to the number of rogue applications developed that were hacking Google Maps.

Lars Rasmussen, the original developer of Google Maps commented how much he learned from the developer community by watching how they hacked the application in real-time, and they took what they learned and applied it to the API we know today.

With the official launch of the Google Maps API, the trend of API mashups really took off, and there are now over 2000 mashups based on this API. The API demonstrates not only the incredible value of geographic data and mapping APIs, but the power that users can

---

[6] http://goo.gl/G4dq3

have in influencing the direction an application or API takes.

APIs Unlock The Cloud

While Amazon was once thought of simply as an online retailer, its own exploration into computing resources helped push the company -- and its API offerings -- to the next level, demonstrating that you can deliver not just data, but computing infrastructure via an API.

Amazon S3

In March 2006, Amazon launched a new web service. This was a new endeavor for Amazon: a storage web service called Amazon S3.[7]

Amazon S3 or Simple Storage Service was initially just an API. There was no Web interface or mobile app. It was just a RESTful API allowing PUT and GET requests with objects or files.

Developers using the Amazon S3 API were charged $0.15 a gigabyte per month for storing files in the cloud.

With this new approach to delivering an API and a pay as you go billing model, Amazon ushered in a new type of computing we now know as cloud computing.

---

[7] http://goo.gl/B9jfg

Amazon EC2

Five months later, Amazon launched another cloud computing service dubbed Amazon EC2 or Elastic Compute Cloud.[8] Just like its predecessor Amazon S3, Amazon EC2 was just a REST API.

Amazon EC2, combined with Amazon S3 has provided the platform for the next generation of computing with APIs at the core.

Cloud computing showed that APIs have come of age and weren't just for simple data exchanges, APIs were ready to deliver the next generation of the Internet.

Amazon Web Services brought in $500 million in revenue in 2010 and is estimated to be $750 million in 2011. By 2014, it will bring in about $2.5 billion in revenue.[9]

APIs & The Mobile Internet

The next generation of computing won't occur on desktops or even laptops. It will happen from the hands of the masses on mobile smart-phones.

Mobile devices are the fastest growing consumer of Web APIs. These APIs provide the data and functionality that drive mobile applications and enable users' mobile browsing and mobile app experiences.

---

[8] http://goo.gl/ElrA1
[9] http://goo.gl/hnoM2

Foursquare

In March 2009 Foursquare launched at the SXSW interactive festival in Austin, Texas.[10]

Foursquare is a location-based mobile platform that makes cities more interesting to explore. By checking in via a smart-phone app or SMS, users share their location with friends while collecting points and earning virtual badges.

In November 2009, Foursquare launched its API to the public. By that time, Foursquare had an impressive set of applications developed by a closed group of partners, but opening the API meant that others could develop apps using the location and check-in mechanism.

Instagram

Instagram launched its photo-sharing iPhone application in October 2010, and less than three months later, boasted over one million users.[11] A powerful, but simple iPhone app, Instagram solved common problems with the quality of mobile photos as well as users' frustrations with sharing.

---

[10] http://goo.gl/qmaWS
[11] http://goo.gl/20veb

However, many users complained about the lack of central Instagram web site and developers lamented the lack of an API. In December a developer name Mislav Marohni□ took it upon himself to reverse engineer the iPhone app and built his own unofficial Instagram API.[12]

Not surprisingly, Instagram shut down the rogue API but then announced it was building one of its own.[13] In February of 2011, Instagram released the official API.[14] Within days many photo applications, photo-sharing sites, and mashups using the Instagram API started to appear.

While Instagram had quickly become a viral sensation as an iPhone app, the availability of an API allowed it to expand even further.

---

[12] http://goo.gl/wC8vu
[13] http://goo.gl/ARnmL
[14] http://goo.gl/XnS4X

This History Is Only the Beginning...

A lot of innovation has occurred within the API space in the last decade, and despite the success of many of the services discussed in this chapter, there have been many fumbles along the way.

As Web APIs are gaining wider acceptance, it's important to look at some of these stumbling blocks and to learn from mistakes that other developers and companies have made when implementing an API.

6
Where Some APIs Stumble

Despite all the successes these API pioneers have had, they've also run into their share of obstacles along the way. Some of these problems were unforeseen, but others were caused by the API owners themselves. As such, we can learn a lot from studying their history, and we can learn from their mistakes.

When building, deploying, and managing APIs, there are plenty of mistakes you can make. The most common ones seem to involve failing to provide clear documentation, overlooking the importance of developer support, and not having a clear road-map for the API and its ecosystem.

Quality API Docs & Resources

Documentation may be the most important key to opening the door of your API, and the most common mistake an API owner can make is to not provide simple and complete API documentation. Without good quality FAQs, tutorials, code samples, case studies, articles and other resources, developers will never understand how to integrate with an API.

Good technical documentation is not easy or cheap.

Reference documentation needs to start with a simple outline of all your API endpoints. Show developers what it can do.

The documentation needs to balance simplicity and completeness, and supporting materials need to be well-organized so that developers can search and navigate them.

Developers do not have the time to spend hours understanding an API. Incomplete or overly complicated documentation can run off even the most seasoned developer. The goal is to get any developer up and running quickly, ready to hack on your API.

No Developer Spotlight

Developers need to be the focus of your API. They will make or break your API community.

Developers need the spotlight on them -- not always in a public way, of course -- but they do need the focus of your attention and resources.

However not all developers will approach your API the same way, and you might find yourself with different strategies to help support the various developer communities.

Partner Developers

Your partner developers, for example, may be the cornerstone of your API initiative. As such, these developers should be given a lot of attention when it comes to your API area.

Partner developers will be motivated by and beholden to their companies collective interests. As they are paid by their company to make the API integration happen, the success of the project lies in both companies interest.

Freelance Developers

Freelance developers have different motives than your partners will. Freelance developers also want to have successful projects, of course, and they're motivated by the potential for financial gain. But many are also in search of notoriety or some sort of developer fame. It is common for developers to take on API development purely for the challenge and for recognition for their work.

There are a number of things you can do to provide exposure and recognition for freelance developers, including contests, hackathons, profile pages, and application showcases.

Even though these freelance developers will utilize a lot of the same resources as partner developers, they may need radically different things when it comes to the "spotlight" and they may have different demands for the attention of your company.

It can be difficult to manage an army of developers, so breaking them into groups and identifying their specific needs is important. This sort of segmentation will enable you to target the support and resources necessary for each group.

No Developer Voice

Providing a way to showcase your developers is important, but you need to do more than just shine a spotlight on successful developers. You need to give your developer community a voice.

Doing so starts by providing a support form or email address. Next, forums provide your developers with not only an individual voice, but a collective community voice.

In order to give developers a way to speak throughout your platform, you can encourage them to share ideas, submit code samples, submit and fork code repositories, showcase their applications, and comment on the API road-map. Your developers can play a key role in research and development around your API, and giving them a true voice can actually contribute ideas, code, and resources to your API.

The more you empower your developers to be involved, the more they will own and help guide your API community. Your developers may not not understand the objectives of your company, but they will understand how to use your use your API.

No Revenue Strategy

An API has to have a revenue strategy that assesses the value the API delivers to end-users. Even if your goal is to offer an API for free and use as a marketing tool, you need a measurement of what this costs and what value its returning to an organization.

You have to ask: does the value outweigh the cost? And where do other players -- partners and developers, for example -- fit into your revenue strategy? Are there opportunities to share costs and share revenue and build an economy within the API community?

It's crucial to think through these things ahead of time because changing the terms of use of an API and/or monetizing aspects that were formerly held by API developers will send a very damaging message to the developer community.

Creating and supporting ways developers and partners can monetize their applications will not only help them be successful and keep using an API, it actually is the first step on a road to building a self-sustaining economy.

7
Building Blocks for a Successful API

While the most popular and successful APIs available today serve a variety of industry needs -- e-commerce, cloud computing, social media, mobile technologies, and so on -- there are some commonalities that many share. There are certain tools and systems in place that are being used across the board, providing a foundation for partners and developers to help establish an API community.

API Overview

An API overview may be the simplest, and yet the most commonly overlooked building block for an API. Rather than launching into the technical details of your API, an overview should tell the story of your API: why it was created and the problems it will solve.

An overview is just that -- something that can be read and understood quickly, so that even the most non-technical user understands what the API does. Remember: it isn't just a developer who's likely to find your API, so you need to keep the overview understandable. Technical details can go elsewhere.

Getting Started

Simply getting started with an API can be the hardest part, as countless hours can be spent fumbling around, trying to understand where to begin. How do I register? Are there code samples in my preferred programming language? How can I get support?

A "Getting Started" page should be easily accessible, ideally following the Overview. This page should have all the pieces of information a developer needs to understand what the API does and how to start to use it.

API Reference / Documentation

An API's documentation will be the map developers use to discover where things are located and how things work. If the map isn't complete, developers will get lost.

The documentation also needs to be easy-to-use and up-to-date. It should be treated as a living document, and the documentation should change as your API evolves.

Code Samples

An overview and good documentation can give a developer a solid introduction to the API, but the move from plan to production can be greatly facilitated by providing as many usable pieces of code as possible.

These needn't all be created in-house, and you can hire outside developers to help build out your code libraries. Again, listen to your developer community to identify what new types of code will make their integration go more quickly and be more successful.

As code samples are created, make sure they're licensed as openly as possible, empowering developers to freely use, distribute, evolve and contribute code back to the API community.

Blog

Blogs are well-established as a communication tool for businesses to highlight the company, the product, the people. Many businesses have also started separate engineering blogs where they can make more technical announcements.

A blog can also be a primary voice to communicate to and about the API community, and RSS syndication can help distribute this message. As with any blog, of course, keeping it active and organized, reflecting what is happening within your API ecosystem, will be an asset, while an inactive blog can quickly become a liability.

Forum

A forum or online group is one of the most common building blocks of an API community. It is a proven part of keeping your API developers and partners educated, interacting and supporting each other.

Forums provide a tool to manage large groups in a format that is familiar and known to a wide number of Internet users. If your forum is public, it can also provide some great SEO value to the online marketing efforts of an API community. If a user is searching at Google for solutions to their problem, a post on an API forum may have the answer they are looking for.

FAQ

An API community will have questions, and users have to find answers as quickly as possible. Providing a self-service system in the form of an FAQ will make it easy for developers to easily find the answers to common questions.

Of course, an FAQ needs to be a living system so it truly reflects the frequently asked questions that are being received via email, social streams and support tickets, and you'll want to provide a way for developers to submit questions for eventual inclusion.

Pricing

Developers need to be able to understand what they are getting and what it's going to cost them. And even if an API is offered for free, there needs to be a clear establishment of pricing. This clarity will set expectations with the community.

Terms of Service

A business must define the terms and conditions under which others can use the API. Developers should have an understanding of is expected when consuming an API, particularly when the API integration is part of their business operations.
Terms of use should be revisted regularly and additions should be considered as part of an ongoing API management strategy. If new endpoints are added or

new types of data opened up, how it impacts the terms and conditions should be evaluated.

Make your terms of use clear and readable by the average person, but make sure they also pass with the lawyers.

Self-Registration

In the spirit of getting your developers up-and-running as painlessly as possible, you want to be able to offer self-service registration. In other words, a developer needs to be able to quickly sign up to use the API and get an API key in return, without any delay or intervention from your staff.

Even if your company requires that developers be approved, you can consider a system by which developers can make a limited number of calls against an API before approval. This gives developers a chance to test-drive the API.

The Basic Building Blocks Aren't Always Enough

If you take a look at the most successful APIs, they will have these common building blocks. But this doesn't mean there isn't room for innovation.

The Future of API Communities: Innovative Building Blocks

As APIs become more important and commonplace, many businesses are finding new and innovative ways to expand their API usage and communities. Much of these efforts involve emphasizing developers and helping make them successful.

Application Gallery

Any application built on an API can be showcased. A gallery of apps demonstrates that people are using the API and are finding success. A gallery can also give developers ideas based upon what others are doing, and it can show new developers that there is a community around the API.

Featured Developer

Although an application gallery showcases what developers have built, it's important that not just their work but actually their profiles be showcased. Developers are the center of any healthy API ecosystem. This can be as simple as creating a "featured developer" section of your site, providing names, pictures, and potentially detailed descriptions of developers.

Hire a Developer

In addition to featuring developers and their applications, you can provide a way for developers to list themselves and their skills so that any business looking for a programmer to work with the API can easily identify someone who has experience.

Ideas

Hopefully APIs owners have an abundance of ideas on how to use their API, ideas that the company cannot find time or resources to do. By building an idea management system, API service providers can share with those in the developer community ideas that someone else can run with.

An "Ideas" section may help a new user or partner understand what an API does and the business value it delivers. A single, published idea might speak to a potential user. And most importantly, Ideas can flourish and grow as they are shared and discussed.

Hackathons

Hackathons bring developers together -- either face-to-face or virtually -- to hack in groups to create some application or code examples that meet the focus and goals of the event.

Hackathons bring together some of the best parts of being a hacker: passion, talent, and ideas (plus caffeine) and can be an environment where innovation happens. Hackathons are becoming a popular way to raise awareness and encourage development around an API. They are also becoming arenas for identifying potential tech talent and potential investment ideas.

Contests

Contests also bring attention to an API. By giving away cash and prizes, you can encourage developers to try out your API. It's also an occasion for press coverage for the community.

twilio contests

Save the Planet With Twilio

Earth Day is just around the corner (April 22nd), so we couldn't help but think how we could help. And what better way to do that than with an army of planeteers—I mean engineers? For this week's contest, we want to see how you can use Twilio to make the Earth a better place. Use Twilio to coordinate a clean-up event or a system that helps people find the **closest recycling center**. Get creative, we want to see how our developer community uses Twilio to have a positive impact on our planet!

Building an Ecosystem to Attract Developers

Competition in the APIs space is growing significantly. Innovation is required to get the attention of developers. Innovative approaches to your API community can bring passionate and creative developers into it. These developers can potentially become champions for your API and lead other developers to integrate with it.

8
Fostering an API Ecosystem

An API starts with the desire to share data or resources that a company offers. It's built with technologies like REST, XML, and JSON, and supported through documentation, along with a handful of code samples to show how to use it.

An API and its supporting developer area are created. And then, what's next? How does a simple API area build community? How does it evolve into a thriving ecosystem like Facebook's of Foursquare's?

It all starts with developers. Giving developers a self-service, resource-rich environment where they have the spotlight and a voice will encourage them in turn to contribute to the API community.

An API owner has to support its API's community, be proactive about reaching out to its community and know what it needs. API support needs to be available in basic forms like a forum, but also there needs to be ways for partners and developers to pay for support and receive premium attention as well.

Resources for developers need to be abundant and well organized. Common resources like blogs, forum, and FAQs are necessary. Tutorials, case studies and "How To's" can take things even further.

Support and resources can create a positive feedback loop among developers and encourage activity that will ideally spread to other users.

Developers can't be expected to visit an API area regularly, so an API community needs to extend its reach to existing social network and developer communities including Twitter, LinkedIn, Github, and Stack Exchange.

Even more than just a presence on these social networks, an API needs to have an offline presence too -- something that can be accomplished by attending conferences, meet-ups, and hackathons, for example. These activities will only serve to strengthen the API community.

The Twilio API

Twilio is a great example of a thriving API. Twilio provides a Web API for business to build scalable voice and SMS applications. Twilio applies common building blocks I've laid out in Chapter 7 and actively innovates in ways that grow, support and empower its development community. Twilio has a ubiquitous presence in Silicon Valley and beyond, establishing partnerships with other API owners and service providers.

The Twilio API ecosystem also provides revenue opportunities for its community through jobs and contests. Twilio goes one step farther than just creating a "jobs board," and by partnering with early stage investment fund 500 Startups[15] to provide investment opportunities for developers' projects.

---

[15] http://goo.gl/Uny2j

9
BizDev 2.5

In the last five years since the introduction of the term "BizDev 2.0," Web APIs have evolved, having being applied to social networking, cloud computing, mobile and beyond.

APIs empower businesses to build platforms, deploy infrastructure, engage in social media marketing, and reach customers via mobile devices. These APIs are developing new opportunities for businesses to grow in a truly global market. As the number of APIs grow and innovation continues, we can safely say we are moving toward BizDev to 2.5.

APIs open revenue opportunities for businesses, their partners and developers -- all of whom build upon the API. With a proper monetization strategy that considers everyone involved in an API community, a company's API feeds its partners and developers, and those developers and partners in turn feed back into the ecosystem.

APIs not only open up new revenue opportunities, they can create research and development environments where developers can innovate in ways never conceived by the original API developers.

Open APIs, hackathons, and idea showcases can identify developer talent and innovative applications that can take give a company the competitive advantage needed to lead.

With the proper business and financial investment an API can become an open R&D ecosystem taking business development to the next level.

Kin Lane

Kin Lane is the API Evangelist for Mimeo.com, he brings unique blend of a IT, data, programming, product development, business development, online and social media marketing talent to the print industry via the Silicon Valley.

He spends his days assisting application developers understand what is possible with the next generation of print and studying best practices when it comes to the business of APIs.

twitter: @kinlane
web: apievangelist.com

Audrey Watters

Audrey Watters is a technology writer and rabble-rouser with a Master's Degree in Folklore. She was working on a PhD in Comparative Literature but chose instead to write about data, culture, education and technology outside of academia rather than finish her dissertation. Audrey has written for ReadWriteWeb, The Huffington Post, O'Reilly Radar, and NPR.

twitter: @audreywatters
web: hackeducation.com

Made in the USA
San Bernardino, CA
13 June 2013